PLATONIC
& ARCHIMEDEAN
SOLIDS

Originally published in Wales by Wooden Books Ltd. in 1998; first published in the United States of America in 2002 by Walker Publishing Company, Inc.

Published simultaneously in Canada by Fitzhenry and Whiteside, Markham, Ontario L3R 4T8

Printed on recycled paper.

For information about permission to reproduce selections from this book, write to Permissions, Walker & Company, 435 Hudson Street, New York, New York 10014

Library of Congress Cataloging-in-Publication Data

Sutton, Daud.
    Platonic & Archimedean solids/written and illustrated by Daud Sutton.
    p.cm.
    Originally published: Presteigne, Powys, Wales : Wooden Books, 1998.
    ISBN 0-8027-1386-6 (alk. paper)
    1. Polyhedra. I. Title: Platonic and Archimedean solids. II. Title.

QA491 .S88 2002
516'.15—dc21

                                                                2001055924

Visit Walker & Company's Web site at www.walkerbooks.com

Printed in the United States of America

2   4   6   8   10   9   7   5   3   1

PLATONIC
& ARCHIMEDEAN
SOLIDS

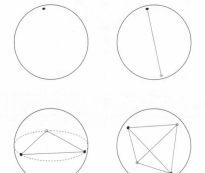

written and illustrated by

Daud Sutton

Walker & Company
New York

In The Name of God,
The Compassionate, The Merciful

This book is dedicated to Professor Keith Critchlow,
whose teaching made it possible.

I am indebted to the many geometers, authors, and
artists who have explored the world of polyhedra.

Thanks to my family and friends
for comments and contributions.

CONTENTS

INTRODUCTION

Imagine a sphere.

It is unity's perfect symbol. Each point on its surface is identical to every other, equidistant from the unique point at its center.

Establishing a single point on the sphere allows others to be defined in relation to it. The simplest and most obvious relationship is with the point directly opposite, found by extending a line through the sphere's center to the other side. Add a third point and space all three as far from each other as possible to define an equilateral triangle. The three points lie on a circle with a radius equal to the sphere's and sharing its center, an example of the largest circles possible on a sphere, known as great circles. Point, line, and triangle occupy zero, one, and two dimensions respectively. It takes a minimum of four points to define an uncurved three-dimensional form.

This small book charts the unfolding of number in three-dimensional space through the most fundamental forms derived from the sphere. A cornerstone of mathematical and artistic inquiry since antiquity, after countless generations these beautiful forms continue to intrigue and inspire.

Cairo, Summer 2001

THE PLATONIC SOLIDS
beautiful forms unfold from unity

Imagine you are on a desert island; there are sticks and sheets of bark. If you start experimenting with making three-dimensional structures you may well discover five "perfect" shapes. In each case they look the same from any *vertex* (corner point), their faces are all made of the same regular shape, and every edge is identical. Their vertices are the most symmetrical distributions of four, six, eight, twelve, and twenty points on a sphere (*below*).

These forms are examples of *polyhedra*, literally "many seats," and, as the earliest surviving description of them as a group is in Plato's *Timaeus*, they are often called the Platonic solids. Plato lived from 427 B.C. to 347 B.C., but there is evidence that they were discovered much earlier (*see page 20*).

The *cube*, with its six square faces, is well known. The other four have names deriving from their numbers of faces. Three of the solids have faces of equilateral triangles: the *tetrahedron* is made from four, the *octahedron* eight, and the *icosahedron* twenty. The *dodecahedron* has twelve regular pentagonal faces. The following ten pages will describe these striking three-dimensional forms in greater detail.

tetrahedron

cube

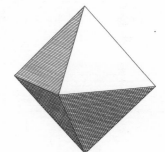

octahedron

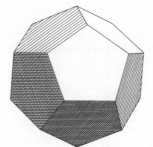

dodecahedron

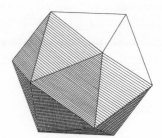

icosahedron

THE TETRAHEDRON
4 faces, 6 edges, 4 vertices

The tetrahedron is composed of four equilateral triangles, with three meeting at every vertex. Its vertices can also be defined by the centers of four touching spheres (*opposite, bottom right*). Plato associated its form with the element of fire because of the penetrating acuteness of its edges and vertices, and because it is the simplest and most fundamental of the regular solids. The Greeks also knew the tetrahedron as *puramis*, from which the word *pyramid* is derived. Curiously the Greek word for fire is *pur*.

The tetrahedron has three 2-fold axes of symmetry, passing through the midpoints of its edges, and four 3-fold axes, each passing through one vertex and the opposite face center (*below*). Any polyhedron with these rotation axes has *tetrahedral symmetry*.

Each Platonic solid is contained by its *circumsphere*, which just touches every vertex. The solids also define two more spheres: their *midsphere*, which passes through the midpoint of every edge, and their *insphere*, which is contained by the solid, perfectly touching the center of every face. For the tetrahedron the *inradius* is one-third of the *circumradius* (*opposite, bottom left*).

edge on : 2-fold   *face on : 3-fold*   *from vertex : 3-fold*

4

THE OCTAHEDRON
8 faces, 12 edges, 6 vertices

The octahedron is made of eight equilateral triangles, four meeting at every vertex. Plato considered the octahedron an intermediary between the tetrahedron, or fire, and the icosahedron, or water and thus ascribed it to the element of air. The octahedron has six 2-fold axes passing through opposite edges, four 3-fold axes passing through its face centers, and three 4-fold axes passing through opposite vertices (*below*). Any polyhedron combining these rotation axes is said to have *octahedral symmetry*.

Greek writings attribute the discovery of the octahedron and icosahedron to Theaetetus of Athens (417 B.C.–369 B.C.). Book XIII of Euclid's *Elements* (*see page 14*) is thought to be based on Theaetetus' work on the regular solids.

The octahedron's circumradius is bigger than its inradius by a factor of $\sqrt{3}$ (*see page 55*). The same relationship occurs between the circumradius and inradius of the cube, and between the circumradius and midradius of the tetrahedron.

The tetrahedron, the octahedron and the cube are all found in the mineral kingdom. Mineral diamonds often form octahedra.

edge on : 2-fold   *face on : 3-fold*   *from vertex : 4-fold*

7

THE ICOSAHEDRON
20 faces, 30 edges, 12 vertices

The icosahedron is composed of twenty equilateral triangles, five to a vertex. It has fifteen 2-fold axes, twenty 3-fold axes, and twelve 5-fold axes (*below*), known as *icosahedral symmetry*. When the tetrahedron, octahedron, and icosahedron are made of identical triangles, the icosahedron is the largest. This led Plato to associate the icosahedron with water, the densest and least penetrating of the three fluid elements—fire, air, and water.

The angle where two faces of a polyhedron meet at an edge is known as a *dihedral angle*. The icosahedron is the Platonic solid with the largest dihedral angles.

If you join the two ends of an icosahedron's edge to the center of the solid an isosceles triangle is defined. This triangle is the same as those that make up the faces of the Great Pyramid at Giza in Egypt.

Arranging twelve equal spheres to define an icosahedron leaves space at the center for another sphere just over nine-tenths as wide as the others (*opposite, lower right*).

edge on : 2-fold

face on : 3-fold

from vertex : 5-fold

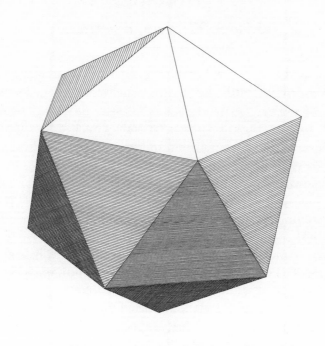

9

THE CUBE
6 faces, 12 edges, 8 vertices

The cube has octahedral symmetry (*below*). Plato assigned it to the element of earth due to the stability of its square bases. Aligned to our experience of space it faces forward, backward, right, left, up, and down, corresponding to the six directions north, south, east, west, zenith, and nadir. Six is the first *perfect number*, with factors adding up to itself ($1 + 2 + 3 = 6$).

Add the cube's twelve edges, the twelve face diagonals, and the four interior diagonals to find a total of twenty-eight straight paths joining the cube's eight vertices to each other. Twenty-eight is the second perfect number ($1 + 2 + 4 + 7 + 14 = 28$).

Islam's annual pilgrimage is to the Kaaba, literally cube, in Mecca. The sanctuary of the Temple of Solomon was a cube, as is the crystalline New Jerusalem in Saint John's revelation. In 430 B.C. the oracle at Delphi instructed the Athenians to double the volume of the cubic altar of Apollo while maintaining its shape. "Doubling the cube," as the problem became known, ultimately proved impossible using Euclidean geometry alone.

edge on : 2-fold

from vertex : 3-fold

face on : 4-fold

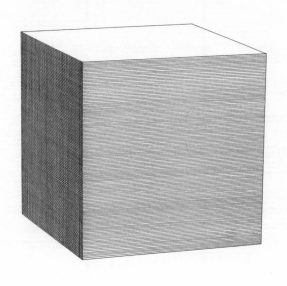

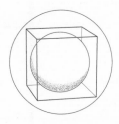

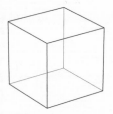

THE DODECAHEDRON
12 faces, 30 edges, 20 vertices

The beautiful dodecahedron has twelve regular pentagonal faces, three of which meet at every vertex. Its symmetry is icosahedral (*below*). Like the tetrahedron, or pyramid, and the cube, the dodecahedron was known to the early Pythagoreans and was commonly referred to as *the sphere of twelve pentagons*. Having detailed the other four solids and ascribed them to the elements, Plato's *Timaeus* says enigmatically, "There remained a fifth construction which God used for embroidering the constellations on the whole heaven."

A dodecahedron sitting on a horizontal surface has vertices lying in four horizontal planes that cut the dodecahedron into three parts. Surprisingly, the middle part is equal in volume to the others, so each is one-third of the total! Also, when set in the same sphere, the surface areas of the icosahedron and dodecahedron are in the same ratio as their volumes.

"Fool's Gold," or iron pyrite, forms crystals much like the dodecahedron, but don't be fooled, their pentagonal faces are not regular and their symmetry is tetrahedral.

edge on : 2-fold   *from vertex : 3-fold*   *face on : 5-fold*

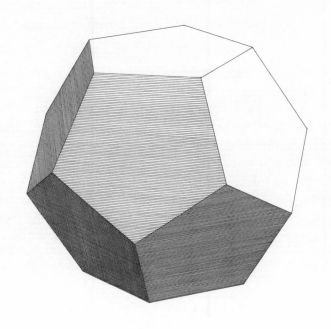

 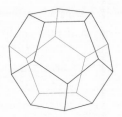 

13

A SHORT PROOF
are there really only five?

A regular polygon has equal sides and angles. A regular polyhedron has equal regular polygon faces and identical vertices. The five Platonic solids are the only convex regular polyhedra.

At least three polygons are needed to make a *solid angle*. Using equilateral triangles this is possible with three (*A*), four (*B*), and five (*C*) around a point. With six the result lies flat (*D*). Three squares make a solid angle (*E*), but with four (*F*) a limit similar to six triangles is reached. Three regular pentagons form a solid angle (*G*), but there is no room, even lying flat, for four or more. Three regular hexagons meeting at a point lie flat (*H*), and higher polygons cannot meet with three around a point, so a final limit is reached. Since only five solid angles made of identical regular polygons are possible, there are at most five possible convex regular polyhedra. Incredibly, all five regular solid angles repeat to form the regular polyhedra. This proof is given by Euclid of Alexandria (c. 325 B.C.–265 B.C.) in Book XIII of his *Elements*.

The angle left as a gap when a polyhedron's vertex is folded flat is its *angle deficiency*. René Descartes (1596–1650) discovered that the sum of a convex polyhedron's angle deficiencies always equals 720°, or two full turns. Later, in the eighteenth century, Leonhard Euler (1707–1783) noticed another peculiar fact: In every convex polyhedron the number of faces minus the number of edges plus the number of vertices equals two.

A

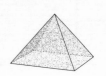

B

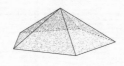

C

D

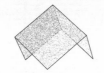

E

F

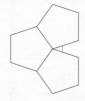
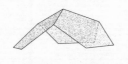

G

H

ALL THINGS IN PAIRS
Platonic solids two by two

What happens if we join the face centers of the Platonic solids? Starting with a tetrahedron, we discover another, inverted, tetrahedron. The faces of a cube produce an octahedron, and an octahedron creates a cube. The icosahedron and dodecahedron likewise produce each other. Two polyhedra whose faces and vertices correspond perfectly are known as each other's *duals*. The tetrahedron is *self-dual*. Dual polyhedra have the same number of edges and the same symmetries.

The illustrations opposite are stereogram pairs. Hold the book at arms length and place a finger vertically, midway to the page. Focus on the finger and then bring the central blurred image into focus. The image should jump into three dimensions.

Dual pairs of Platonic solids can be joined with their edges touching at their midpoints to give the compound polyhedra shown below. Everything in creation has its counterpart or opposite, and the dual relationships of the Platonic solids are a beautiful example of this principle.

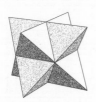 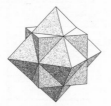 

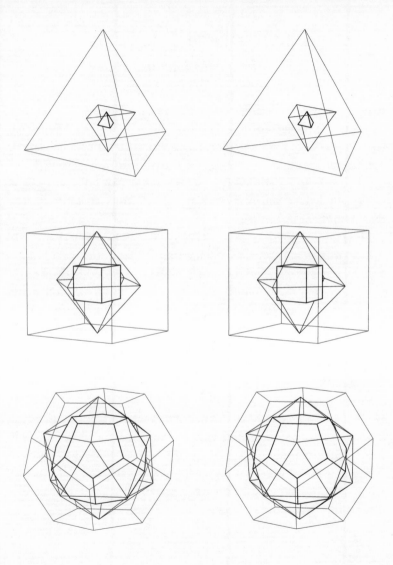

17

AROUND THE GLOBE
in elegant ways

Plato's cosmology constructs the elemental solids from two types of right-triangular atoms. The first atom is half an equilateral triangle, six of which then compound to produce larger equilateral triangles; these go on to form the tetrahedron, octahedron, and icosahedron. The second triangular atom is a diagonally halved square, which appears in fours, making squares that then form cubes.

The Platonic solids have planes of symmetry dividing them into mirror-image halves. The tetrahedron has six, the octahedron and cube have nine, and the icosahedron and dodecahedron have fifteen. When the tetrahedron, octahedron, and icosahedron are constructed from Plato's triangular atoms, paths are defined that make their mirror planes explicit. The cube, however, needs twice as many triangular divisions as Plato gave it (*top row*) to delineate all its mirror planes (*middle row*).

Projecting the subdivided Platonic solids onto their circumspheres produces three spherical systems of symmetry. Each spherical system is defined by a characteristic spherical triangle with one right angle and one angle of one-third of a half turn. Their third angles are respectively one-third of a half turn (*top row*), one-quarter of a half turn (*middle row*), and one-fifth of a half turn (*bottom row*). This sequence of $\frac{1}{3}$, $\frac{1}{4}$, and $\frac{1}{5}$ elegantly inverts the Pythagorean whole number triple 3, 4, 5.

ROUND AND ROUND
lesser circles

Any navigator will tell you that the shortest distance between two points on a sphere's surface is always an arc of a *great circle*. When a polyhedron's edges are projected onto its circumsphere the result is a set of great circle arcs known as a *radial projection*. The left-hand column opposite shows the radial projections of the Platonic solids with their great circles shown as dotted lines.

A spherical circle smaller than a great circle is called a *lesser circle*. Tracing a circle around all the faces of the Platonic solids set in their circumspheres generates the patterns of lesser circles, shown in the middle column. Book XIV of Euclid's *Elements* proves that when set in the same sphere, the lesser circles around the dodecahedron's faces (*fourth row*) are equal to the lesser circles around the icosahedron's faces (*fifth row*). The same is true of the cube (*second row*) and the octahedron (*third row*) as a pair.

Shrink the lesser circles in the middle column until they just touch each other to define the five spherical curiosities in the right-hand column. Many neolithic carved stone spheres have been found in Scotland with the same patterns as the first four of these arrangements. The dodecahedral carvings of twelve circles on a sphere, some 4,000 years old, are the earliest known examples of manmade designs with icosahedral symmetry.

Large lesser circle models can be made from circles of willow, or cheap hula-hoops, lashed together with wire, string, or tape.

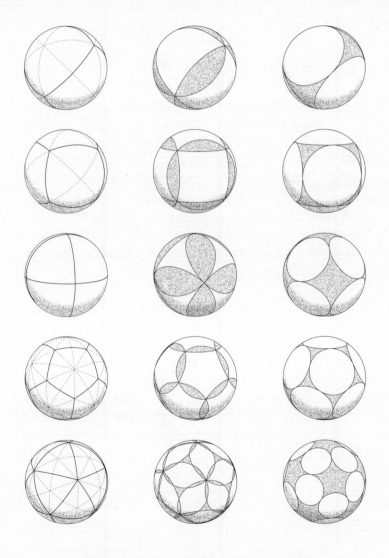

THE GOLDEN RATIO
and some intriguing juxtapositions

Dividing a line so that the shorter section is to the longer as the longer section is to the whole line defines the *golden ratio* (*below*). It is an irrational number, inexpressible as a simple fraction (*see page 55*). Its value is one plus the square root of five, divided by two—approximately 1.618. It is represented by the Greek letter φ (*phi*) or sometimes by τ (*tau*). φ has intimate connections with unity; φ times itself (φ²) is equal to φ plus one (2.618 . . .), and one divided by φ equals φ minus one (0.618 . . .). It is innately related to five-fold symmetry; each successive pair of heavy lines in the pentagram below is in the golden ratio.

A *golden rectangle* has sides in the golden ratio. If a square is removed from one side, the remaining rectangle is another golden rectangle. This process can continue indefinitely and establishes a golden spiral (*below right*). Remarkably, an icosahedron's twelve vertices are defined by three perpendicular golden rectangles (*opposite, top*). The dodecahedron is even richer. Twelve of its twenty vertices are defined by three perpendicular φ² rectangles, and the remaining eight vertices are found by adding a cube of edge length φ (*opposite, bottom*).

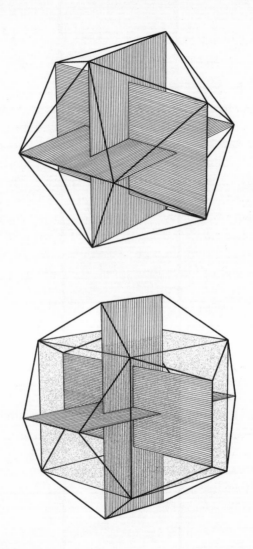

POLYHEDRA WITHIN POLYHEDRA
and so proceed ad infinitum

The Platonic solids fit together in remarkable and fascinating ways. Page 54 shows many of those relationships. The upper stereogram pair opposite shows a dodecahedron with edge length one. Nested inside it is a cube, edge length ϕ, and a tetrahedron, edge length $\sqrt{2}$ (*see page 55*) times the cube's. The tetrahedron occupies one-third of the cube's volume.

In the lower stereogram pair opposite, the six edge midpoints of the tetrahedron define the six vertices of an octahedron. As well as halving the tetrahedron's edges this octahedron has half its surface area and half its volume, perfectly embodying the musical octave ratio of 1:2. Similarly the twelve edges of the octahedron correspond to the twelve vertices of a nested icosahedron. The icosahedron's vertices cut the octahedron's edges perfectly into the golden ratio.

Imagine these two sets of nestings combined to give all five Platonic solids in one elegant arrangement. Since the outer dodecahedron defines a larger icosahedron by their dual relationship, and the inner icosahedron likewise defines a smaller dodecahedron, the nestings can be continued outward and inward to infinity.

The tetrahedron, octahedron, and icosahedron, made entirely from equilateral triangles, are known as *convex deltahedra*, after the Greek letter Δ (*delta*). The five other possible convex deltahedra are shown in the bottom row opposite.

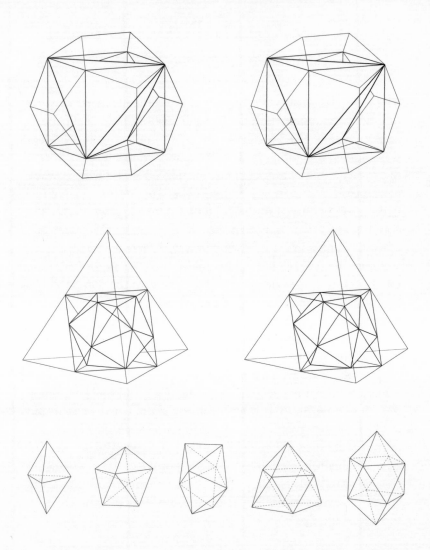

COMPOUND POLYHEDRA
a stretch of the imagination

The interrelationships on the previous page generate particularly beautiful compound polyhedra. Fix the position of an icosahedron, and octahedra can be placed around it in five different ways, giving the compound of five octahedra (*top left*). Similarly the cube within the dodecahedron, placed five different ways, generates the compound of five cubes (*top right*). The tetrahedron can be placed in the cube two different ways to give the compound of two tetrahedra shown on page 16. Replace each of the five cubes in the dodecahedron with two tetrahedra to give the compound of ten tetrahedra (*middle left*). Remove five of the tetrahedra from the compound of ten, to leave the compound of five tetrahedra (*middle right*). This occurs in two versions, right-handed, or *dextro,* and left-handed, or *laevo;* the two versions cannot be superimposed and are described as each other's *enantiomorphs.* Polyhedra or compounds with this property of "handedness" are referred to as *chiral.*

Returning to the cube and dodecahedron, and this time fixing the cube, there are two ways to place the dodecahedron around it. The result of both ways used simultaneously is the compound of two dodecahedra (*bottom left*). In the same way the octahedron and icosahedron pair gives the compound of two icosahedra (*bottom right*). Many other extraordinary compound polyhedra are possible; for example, Bakos's compound of four cubes is shown on the first page of this book.

THE KEPLER POLYHEDRA
the stellated and great stellated dodecahedron

The sides of some polygons can be extended until they meet again; for example, the regular pentagon extends to form a five pointed star, or pentagram (*below*). This process is known as *stellation*. Johannes Kepler (1571–1630) proposed its application to polyhedra, observing the two possibilities of stellation by extending edges, and stellation by extending face planes. Applying the first of these (*below*) to the dodecahedron and icosahedron he discovered the two polyhedra illustrated opposite and named them the larger and smaller icosahedral hedgehogs.

Their modern names, the stellated dodecahedron (*opposite, top*) and the great stellated dodecahedron (*opposite, bottom*), reveal that these polyhedra are also two of the face stellations of the dodecahedron. Each is made of twelve pentagram faces, one with five, the other with three to every vertex. They have icosahedral symmetry.

Although its five sides intersect each other, the pentagram has equal edges and equal angles at its vertices and so can be considered a nonconvex regular polygon. Likewise, these polyhedra can be regarded as nonconvex regular polyhedra.

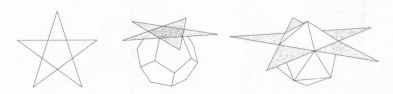

29

THE POINSOT POLYHEDRA
the great dodecahedron and the great icosahedron

Louis Poinsot (1777–1859) investigated polyhedra indepen-
dently of Kepler.   He rediscovered Kepler's two icosahedral
hedgehogs and also discovered the two polyhedra shown here:
the great dodecahedron (*opposite, top*) and the great icosahedron
(*opposite, bottom*).   Both of these polyhedra have five faces to a
vertex, intersecting each other to give pentagram *vertex figures*.
The great dodecahedron has twelve pentagonal faces and is the
third stellation of the dodecahedron. The great icosahedron has
twenty triangular faces and is one of an incredible fifty-nine
possible stellations of the icosahedron, which also include the
compounds of five octahedra and of five and ten tetrahedra.

A nonconvex regular polyhedron must have vertices arranged
like one of the Platonic solids. Joining a polyhedron's vertices to
form new types of polygon within it is known as *faceting*. The
possibilities of faceting the Platonic solids produce the
compounds of two and ten tetrahedra, the compound of five
cubes, the two Poinsot polyhedra (*below left*) and the two Kepler
star polyhedra (*below right*). The four Kepler-Poinsot polyhedra
are therefore the only nonconvex regular polyhedra.

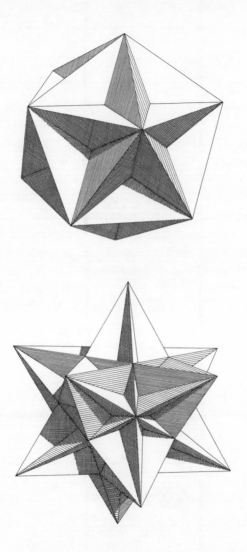

THE ARCHIMEDEAN SOLIDS
thirteen semiregular polyhedra

The thirteen Archimedean solids (*opposite*) are the subject of much of the rest of this book. Also known as the *semiregular polyhedra*, they have regular faces of more than one type, and identical vertices. They all fit perfectly within a sphere, with tetrahedral, octahedral, or icosahedral symmetry. Although their earliest attribution is to Archimedes (c. 287 B.C.–212 B.C.), Kepler seems to have been the first person since antiquity to describe the whole set of thirteen in his *Harmonices Mundi*. He further noted the two infinite sets of regular prisms and antiprisms (*below*), which also have identical vertices and regular faces.

Turn one octagonal cap of the rhombicuboctahedron by an eighth of a turn to obtain the pseudorhombicuboctahedron (*below*). Its vertices, while surrounded by the same regular polygons, are of *two* types relative to the polyhedron as a whole.

There are fifty-three semiregular nonconvex polyhedra, one example being the dodecadodecahedron (*below*). Together with the Platonic and Archimedean solids, and the Kepler-Poinsot polyhedra, they form the set of seventy-five uniform polyhedra.

heptagonal prism   *heptagonal antiprism*   *pseudo rhombicuboctahedron*   *dodecadodecahedron*

truncated tetrahedron

truncated octahedron

cuboctahedron

truncated cube

rhombicuboctahedron

great rhombicuboctahedron

snub cube

truncated icosahedron

icosidodecahedron

truncated dodecahedron

rhombicosidodecahedron

great rhombicosidodecahedron

snub dodecahedron

FIVE TRUNCATIONS
off with their corners!

Truncate the Platonic solids to produce the five equal-edged Archimedean polyhedra shown here. These truncated solids are the perfect demonstration of the Platonic solids' vertex figures: triangular for the tetrahedron, cube, and dodecahedron; square for the octahedron; and pentagonal for the icosahedron. Each Archimedean solid has one circumsphere and one midsphere. They have an insphere for each type of face, the larger faces having the smaller inspheres touching their centers. Each truncated solid therefore defines four concentric spheres.

The five truncated solids can each sit neatly inside both their original Platonic solid and that solid's dual. For example, the truncated cube can rest its octagonal faces within a cube or its triangular faces within an octahedron.

The truncated octahedron is the only Archimedean solid that can fill space with identical copies of itself, leaving no gaps. It also conceals a less obvious secret. Joining the ends of one of its edges to its center produces a central angle that is the same as the acute angle in the famous Pythagorean 3 : 4 : 5 triangle, beloved by ancient Egyptian masons for defining a right angle.

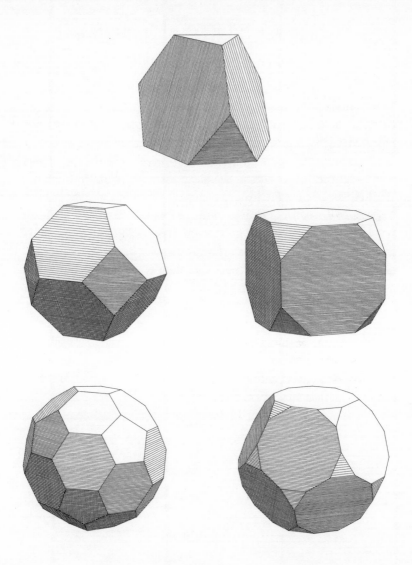

THE CUBOCTAHEDRON
14 faces, 24 edges, 12 vertices

The cuboctahedron combines the six square faces of the cube with the eight triangular faces of the octahedron. It has octahedral symmetry. Joining the edge midpoints of either the cube or the octahedron traces out a cuboctahedron (*shown below as a stereogram pair*). According to Heron of Alexandria (10–75), Archimedes ascribed the cuboctahedron to Plato.

Quasiregular polyhedra such as the cuboctahedron are made of two types of regular polygon, each type being surrounded by polygons of the other type. The identical edges, in addition to defining the faces themselves, also define equatorial polygons. For example, the cuboctahedron's edges define four regular hexagons. The radial projections of quasiregular polyhedra consist entirely of complete great circles (*opposite, bottom left*).

The maximum number of identical spheres that can fit around a central sphere of equal size is twelve. Arranged symmetrically so that their centers define the vertices of a cuboctahedron, they each touch four neighbors (*opposite, bottom right*).

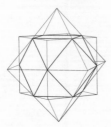 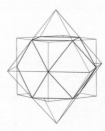

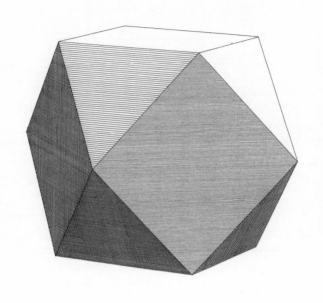

A CUNNING TWIST
and a structural wonder

Picture a cuboctahedron made of rigid struts joined at flexible vertices. This structure was named "the jitterbug" by R. Buckminster Fuller (1895–1983), and is shown opposite with the rigid triangular faces filled in for clarity. The jitterbug can be slowly collapsed in on itself in two ways so that the square "holes" become distorted. When the distance between the closing corners equals the edge length of the triangles, an icosahedron is defined. Continue collapsing the structure and it becomes an octahedron. If the top triangle is then given a twist the structure flattens to form four triangles that close up to give the tetrahedron.

Geodesic domes are another of Buckminster Fuller's structural discoveries. These are parts of geodesic spheres, which are formed by subdividing the faces of a triangular polyhedron, usually the icosahedron, into smaller triangles, and then projecting the new vertices outward to the same distance from the center as the original ones (*below*). A distant relative of the geodesic sphere is the popular Renaissance polyhedron of seventy-two sides known as Campanus's sphere (*below right*).

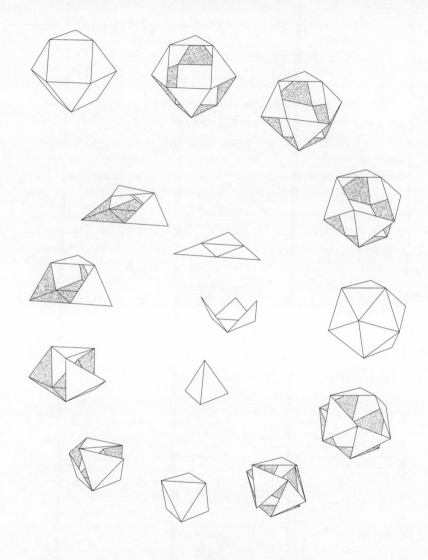

THE ICOSIDODECAHEDRON
32 faces, 60 edges, 30 vertices

The icosidodecahedron combines the twelve pentagonal faces of the dodecahedron with the twenty triangular faces of the icosahedron. Joining the edge midpoints of either the dodecahedron or the icosahedron traces out the quasiregular icosidodecahedron (*both shown below as a stereogram pair*). Its edges form six equatorial decagons, giving a radial projection of six great circles (*opposite, bottom left*).

The earliest known depiction of the icosidodecahedron is by Leonardo Da Vinci (1452–1519) and appears in Fra Luca Pacioli's (1445–1517) *De Divina Proportione*. Appropriately this work's main theme is the golden ratio, which is perfectly embodied by the ratio of the icosidodecahedron's edge to its circumradius.

Defining the icosidodecahedron with thirty equal spheres leaves space for a large central sphere that is √5 (*see page 55*) times as large as the others (*opposite, bottom right*).

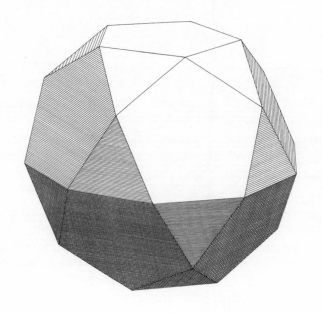

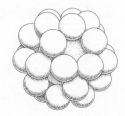

41

FOUR EXPLOSIONS
expanding from the center

Exploding the faces of the cube or the octahedron outward until they are separated by an edge length (*below*) defines the rhombicuboctahedron (*opposite, top left*). The same process applied to the dodecahedron or icosahedron gives the rhombicosidodecahedron (*opposite, top right*). The octagonal faces of the truncated cube, or the hexagonal faces of the truncated octahedron, explode to give the great rhombicuboctahedron (*opposite, bottom left*). The decagonal faces of the truncated dodecahedron, or the hexagonal faces of the truncated icosahedron, explode to give the great rhombicosidodecahedron (*opposite, bottom right*).

Kepler called the great rhombicuboctahedron a truncated cuboctahedron, and the great rhombicosidodecahedron a truncated icosidodecahedron. Truncating these polyhedra, however, does not leave square faces, but √2 and φ rectangles.

These four polyhedra have face planes in common with either the cube, octahedron, and rhombic dodecahedron (*see page 47*), or the icosahedron, dodecahedron, and rhombic triacontahedron (*see page 47*), hence the prefix "rhombi-" in their names.

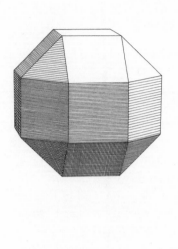

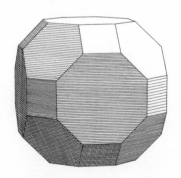

TURNING
the snub cube and the snub dodecahedron

The name "snub cube" is a loose translation of Kepler's name *cubus simus*, literally "the squashed cube." Both the snub cube and the snub dodecahedron are chiral, occurring in dextro and laevo versions. Both versions are illustrated opposite with the dextro versions on the right. The snub cube has octahedral symmetry, and the snub dodecahedron has icosahedral symmetry. Neither has any mirror planes. Of the Platonic and Archimedean solids the snub dodecahedron is closest to the sphere.

The rhombicuboctahedron (*see page 43*) can be used to make a structure similar to the jitterbug (*see page 39*). Applying a twist to this new structure produces the snub cube (*below*). Twist one way to make the dextro version and the other to make the laevo. The corresponding relationship exists between the rhombicosi-dodecahedron and the snub dodecahedron.

The five Platonic solids have been truncated, combined, exploded, and twisted into the thirteen Archimedean solids. Three-dimensional space is revealing its order, complexity, and subtlety. What other wonders await?

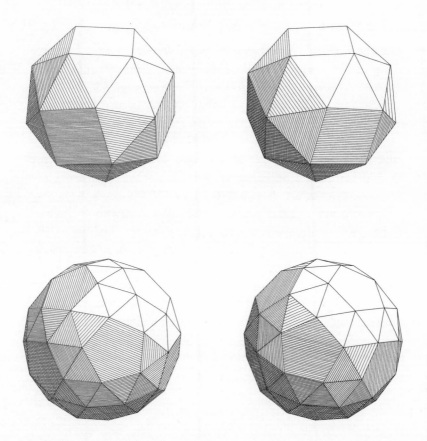

THE ARCHIMEDEAN DUALS
everything has its opposite

The duals of the Archimedean solids were first described as a group by Eugène Catalan (1814–1894) and are positioned opposite to correspond with their partners on page 33. To create the dual of an Archimedean solid, extend perpendicular lines from its edge midpoints, tangential to the solid's midsphere. These lines are the dual's edges, the points where they first intersect each other are its vertices. Archimedean solids have one type of vertex and different types of faces, their duals therefore have one type of face but different types of vertices.

The two quasiregular Archimedean solids, the cuboctahedron and the icosidodecahedron, both have rhombic duals that were discovered by Kepler. The Platonic dual pair compounds (*pages 16, 36, and 40*) define the face diagonals of these rhombic polyhedra, which are in the ratios $\sqrt{2}$ for the rhombic dodecahedron and ϕ for the rhombic triacontahedron. Kepler noticed that bees terminate their hexagonal honeycomb cells with three such $\sqrt{2}$ rhombs. He also described the three dual pairs involving quasiregular solids (*below*), where the cube is seen as a rhombic solid, and the octahedron as a quasiregular solid.

triakistetrahedron

tetrakishexahedron

rhombic dodecahedron

triakisoctahedron

trapezoidal icositetrahedron

disdyakisdodecahedron

pentagonal icositetrahedron

pentakisdodecahedron

rhombic triacontahedron

triakisicosahedron

trapezoidal hexecontahedron

disdyakistriacontahedron

pentagonal hexecontahedron

MORE EXPLOSIONS
and unseen dimensions

Exploding the rhombic dodecahedron, or its dual the cuboctahedron, results in an equal edged convex polyhedron of fifty faces (*opposite, top right*). The exploded rhombic triaconta-hedron, or exploded icosidodecahedron, has one hundred and twenty-two faces (*opposite, bottom right*).

Ludwig Schläfi (1814–1895) proved that there are six regular four-dimensional *polytopes* (generalizations of polyhedra): the 5-cell made of tetrahedra; the 8-cell, or *tesseract,* made of cubes; the 16-cell made of tetrahedra; the 24-cell made of octahedra; the 120-cell made of dodecahedra; and the 600-cell made of tetrahedra. The rhombic dodecahedron is a three-dimensional shadow of the four-dimensional tesseract analogous to the hexagon as a two-dimensional shadow of the cube. In a cube two squares meet at every edge. In a tesseract three squares meet at every edge. Squares through the same edge define three cubes (*shaded below with an alternative tesseract projection*). Schläfi also proved that in five or more dimensions the only regular polytopes are the *simplex*, or generalized tetrahedron, the *hypercube*, or generalized cube, and the *orthoplex*, or generalized octahedron.

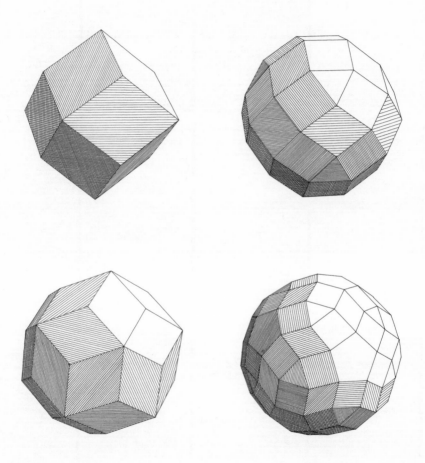

Flat-Packed Polyhedra

If a polyhedron is "undone" along some of its edges and folded flat, the result is known as its *net*. The earliest known examples of polyhedra presented this way are found in Albrecht Dürer's *Painter's Manual,* from 1525. The nets below are scaled such that if refolded the resulting polyhedra would all have equal circumspheres.

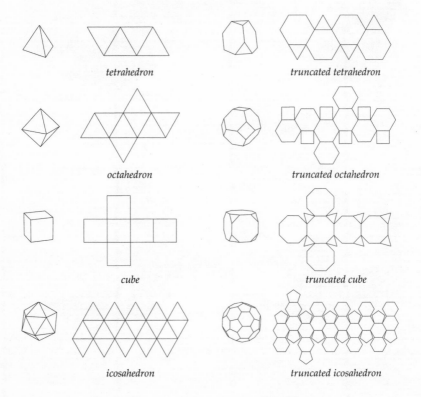

tetrahedron

truncated tetrahedron

octahedron

truncated octahedron

cube

truncated cube

icosahedron

truncated icosahedron

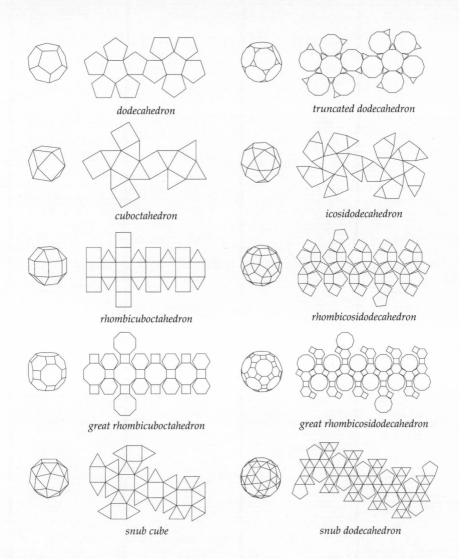

dodecahedron

truncated dodecahedron

cuboctahedron

icosidodecahedron

rhombicuboctahedron

rhombicosidodecahedron

great rhombicuboctahedron

great rhombicosidodecahedron

snub cube

snub dodecahedron

ARCHIMEDEAN SYMMETRIES

The diagrams below show the rotation symmetries of the Archimedean solids and the two rhombic Archimedean duals.

truncated tetrahedron

truncated octahedron

truncated icosahedron

truncated cube

truncated dodecahedron

cuboctahedron

icosidodecahedron

rhombicuboctahedron

rhombicosidodecahedron

great rhombicuboctahedron

great rhombicosidodecahedron

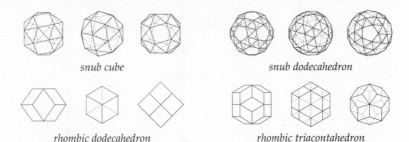

snub cube   *snub dodecahedron*

rhombic dodecahedron   *rhombic triacontahedron*

THREE-DIMENSIONAL TESSELLATIONS

Of the Platonic solids only the cube can fill space with copies of itself and leave no gaps. The only other purely "Platonic" space filling combines tetrahedra and octahedra. One Archimedean solid, the truncated octahedron, and one Archimedean dual, the rhombic dodecahedron, are also space-filling polyhedra.

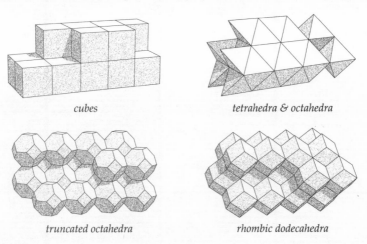

cubes   *tetrahedra & octahedra*

truncated octahedra   *rhombic dodecahedra*

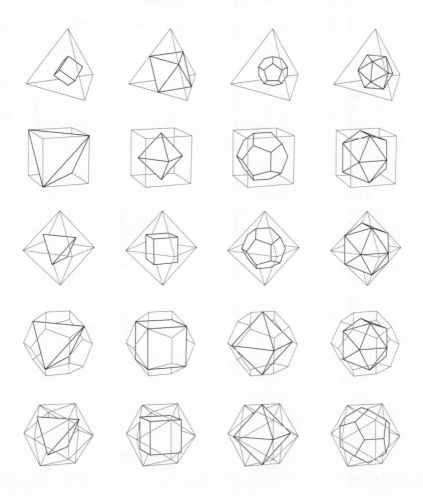

Expansions and Formulas

A recurring theme in the metric properties of the Platonic solids is the occurrence of the irrational numbers phi (ϕ), and the square roots $\sqrt{2}$, $\sqrt{3}$, and $\sqrt{5}$. They are surprisingly elegant when expressed as (infinitely) continued fractions:

$$\phi = 1+\cfrac{1}{1+\cfrac{1}{1+\cfrac{1}{1+\cdots}}} \qquad \sqrt{2} = 1+\cfrac{1}{2+\cfrac{1}{2+\cfrac{1}{2+\cdots}}} \qquad \sqrt{3} = 1+\cfrac{1}{1+\cfrac{1}{2+\cfrac{1}{2+\cdots}}} \qquad \sqrt{5} = 2+\cfrac{1}{4+\cfrac{1}{4+\cfrac{1}{4+\cdots}}}$$

Their decimal expansions to twelve places, together with that of π are

$$\phi = 1.618033988750 \qquad \sqrt{2} = 1.414213562373 \qquad \sqrt{3} = 1.732050807569$$
$$\sqrt{5} = 2.236067977500 \qquad \pi = 3.141592653590$$

The table below gives volumes and surface areas for a sphere radius r, and Platonic solids, edge length s. Also included are the proportional pathways joining each vertex to every other in the Platonic solids.

	Volume	Surface Area	Number of Pathways, Length
Sphere	$\frac{4}{3}\pi r^3$	$4\pi r^2$	n/a
Tetrahedron	$\frac{\sqrt{2}}{12} s^3$	$\sqrt{3}\,s^2$	6 edges, s
Octahedron	$\frac{\sqrt{2}}{3} s^3$	$2\sqrt{3}\,s^2$	12 edges, s 3 axial diagonals, $\sqrt{2}\,s$
Cube	s^3	$6s^2$	12 edges, s 12 face diagonals (inscribed tetrahedra), $\sqrt{2}\,s$ 4 axial diagonals, $\sqrt{3}\,s$
Icosahedron	$\frac{5}{6}\phi^2 s^3$	$5\sqrt{3}\,s^2$	30 edges, s 30 face diagonals, ϕs 6 axial diagonals, $\sqrt{(\phi^2+1)}\,s$
Dodecahedron	$\frac{\sqrt{5}}{2}\phi^4 s^3$	$3\sqrt{(25+10\sqrt{5})}\,s^2$	30 edges, s 60 face diagonals (inscribed cubes), ϕs 60 interior diagonals (inscr. tetrahedra), $\sqrt{2}\phi s$ 30 interior diagonals, $\phi^2 s$ 10 axial diagonals, $\sqrt{3}\phi s$

	Symmetry*	Vertices	Edges	Faces (total)	Faces (types)
DATA					
TABLE Tetrahedron	Tetr.	4	6	4	4 triangles
Cube	Oct.	8	12	6	6 squares
Octahedron	Oct.	6	12	8	8 triangles
Dodecahedron	Icos.	20	30	12	12 pentagons
Icosahedron	Icos.	12	30	20	20 triangles
Stellated Dodecahedron	Icos.	12	30	12	12 pentagrams
Great Dodecahedron	Icos.	12	30	12	12 pentagons
Great Stellated Dodecahedron	Icos.	20	30	12	12 pentagrams
Great Icosahedron	Icos.	12	30	20	20 triangles
Cuboctahedron	Oct.	12	24	14	8 triangles 6 squares
Icosidodecahedron	Icos.	30	60	32	20 triangles 12 pentagons
Truncated Tetrahedron	Tetr.	12	18	8	4 triangles 4 hexagons
Truncated Cube	Oct.	24	36	14	8 triangles 6 octagons
Truncated Octahedron	Oct.	24	36	14	6 squares 8 hexagons
Truncated Dodecahedron	Icos.	60	90	32	20 triangles 12 decagons
Truncated Icosahedron	Icos.	60	90	32	12 pentagons 20 hexagons
Rhombicuboctahedron	Oct.	24	48	26	8 triangles 18 squares
Great Rhombicuboctahedron	Oct.	48	72	26	12 squares 8 hexagons 6 octagons
Rhombicosidodecahedron	Icos.	60	120	62	20 triangles 30 squares 12 pentagons
Great Rhombicosidodecahedron	Icos.	120	180	62	30 squares 20 hexagons 12 decagons
Snub Cube	Oct.-**	24	60	38	32 triangles 6 squares
Snub Dodecahedron	Icos.-**	60	150	92	80 triangles 12 pentagons

* Symmetries: Tetrahedral: 4 x 3-fold axes, 3 x 2-fold, 6 mirror planes.   Octahedral: 3 x 4-fold axes, 4 x 3-fold, 6 x 2-fold, 9 mirror planes.
Icosahedral: 6 x 5-fold axes, 10 x 3-fold, 15 x 2-fold, 15 mirror planes.
** The snub solids have no mirror planes.

Inradius*** / Circumradius	Midradius*** / Circumradius	Edge Length*** / Circumradius	Dihedral Angles****	Central Angle*****
0.3333333333	0.5773502692	1.6329931619	70°31'44"	109°28'16"
0.5773502692	0.8164965809	1.1547005384	90°00'00"	70°31'44"
0.5773502692	0.7071067812	1.4142135624	109°28'16"	90°00'00"
0.7946544723	0.9341723590	0.7136441795	116°33'54"	41°48'37"
0.7946544723	0.8506508084	1.0514622242	138°11'23"	63°26'06"
0.4472135955	0.5257311121	1.7013016167	116°33'54"	116°33'54"
0.4472135955	0.8506508084	1.0514622242	63°26'06"	63°26'06"
0.1875924741	0.3568220898	1.8683447179	63°26'06"	138°11'23"
0.1875924741	0.5257311121	1.7013016167	41°48'37"	116°33'54"
0.8164965809 0.7071067812	0.8660254038	1.0000000000	125°15'52"	60°00'00"
0.9341723590 0.8506508084	0.9510565163	0.6180339887	142°37'21"	36°00'00"
0.8703882798 0.5222329679	0.9045340337	0.8528028654	70°31'44" 109°28'16"	50°28'44"
0.9458621650 0.6785983445	0.9596829823	0.5621692754	90°00'00" 125°15'52"	32°39'00"
0.8944271910 0.7745966692	0.9486832981	0.6324555320	109°28'16" 125°15'52"	36°52'12"
0.9809163757 0.8385051474	0.9857219193	0.3367628118	116°33'54" 142°37'21"	19°23'15"
0.9392336205 0.9149583817	0.9794320855	0.4035482123	138°11'23" 142°37'21"	23°16'53"
0.9108680249 0.8628562095	0.9339488311	0.7148134887	135°00'00" 144°44'08"	41°52'55"
0.9523198087 0.9021230715 0.8259425910	0.9764509762	0.4314788105	125°15'52" 135°00'00" 144°44'08"	24°55'04"
0.9659953695 0.9485360199 0.9245941063	0.9746077624	0.4478379596	148°16'57" 153°56'33" 159°05'41"	25°52'43"
0.9825566436 0.9647979663 0.9049441875	0.9913166895	0.2629921751	142°37'21" 148°16'57" 159°05'41"	15°06'44"
0.9029870683 0.8503402074	0.9281913780	0.7442063312	142°59'00" 153°14'05"	43°41'27"
0.9634723304 0.9188614921	0.9727328506	0.4638568806	152°55'48" 164°10'31"	26°49'17"

*** From the polyhedron's center the inradius is measured to the various face-centers, the midradius to the edge midpoints, and the circumradius to vertices.
**** In Archimedean solids the larger dihedral angles are found between smaller pairs of faces.
***** The central angle is the angle formed at the center of a polyhedron by joining the ends of an edge to that center.

FURTHER READING

If you have enjoyed this Wooden Book, others in the series that may be of interest include *Sacred Geometry* by Miranda Lundy and *Useful Mathematical & Physical Formulæ* by Matthew Watkins.

For those looking for more things polyhedral, Keith Critchlow's *Order In Space* (Thames & Hudson) and Peter R. Cromwell's *Polyhedra* (Cambridge) are both highly recommended. H. S. M. Coxeter's *Regular Polytopes* (Dover) is the classic twentieth-century mathematical text on the subject, and Norman Johnson's forthcoming *Uniform Polytopes* (Cambridge) promises to become an indispensable addition to the literature. Those with access to a manuscript library are well advised to seek out Wenzel Jamnitzer's *Perspectiva Corporum Regularium* (1568).

For those wishing to make models, Magnus J. Wenninger's *Polyhedron Models* (Cambridge), *Dual Models* (Cambridge), and *Spherical Models* (Dover) cover their respective areas very thoroughly. *Shapes, Space and Symmetry* by Alan Holden (Dover) is also good. A range of cut-out-and-make-polyhedra books is published by Tarquin Books.

George Hart's excellent online *Encyclopedia of Polyhedra* contains over 1,000 virtual reality polyhedra, with many accompanying articles and links. It can be found at www.georgehart.com.